A GUI[illegible]ANCY

A short [illegible]tting up a consultancy practice

Second Edition

Second Edition: Rolf Clayton
First Edition: Henry Rowson and Jim Madden

INSTITUTION OF CHEMICAL ENGINEERS

Published by
Institution of Chemical Engineers,
Davis Building,
165–189 Railway Terrace,
Rugby, Warwickshire CV21 3HQ, UK.

ISBN 0 85295 340 2

Printed in the United Kingdom by Chameleon Press Ltd, 5–25 Burr Road, Wandsworth, London SW18 4SG.

FOREWORD TO SECOND EDITION

This edition of the guide has been brought up to date and extended. It includes new sections with practical advice about how to set up an office for your consultancy and information on obtaining help and advice which was accurate at 1 January 1995. It still maintains, however, its emphasis on not ducking the crucial challenges of making a career as a consultant.

FOREWORD TO FIRST EDITION

In response to enquiries from many members in recent years the Institution has prepared this booklet on consultancy. Its purpose is to indicate to members some of the problems as well as some of the opportunities they will encounter should they decide to take on consultancy work. It is definitely *not* a manual for consultants, rather an *aide-mémoire* for those new to the field.

The booklet has drawn heavily on a similar publication issued by the Institution of Mechanical Engineers, which in turn was based on an earlier pamphlet prepared by the Royal Society of Chemistry. We take the opportunity to thank both organizations for their kind permission to use their work so extensively.

CONTENTS

1. INTRODUCTION

THE PURPOSE OF THIS GUIDE

This guide is aimed principally at the individual or group of individuals who are considering establishing themselves as consultants and offering a service by selling the skills and knowledge they possess.

The Institution of Chemical Engineers (IChemE) *List of Consultants 1994/5* lists 98 UK consultancies. There are also a number of other consultancies not listed but which are run by chartered members of the IChemE, many of which are single-person firms. The size distribution of the consultancies listed by the IChemE is given in Table 1.

TABLE 1
Size distribution of consultancies in the IChemE *List of Consultants 1994/5*

Size of consultancy (no. of staff)	Number of consultancies	Percentage
1	43	42%
2	11	11%
3	3	3%
4	3	3%
5	2	2%
6–10	3	3%
11–20	7	7%
21–50	11	11%
51–100	8	8%
>100	7	7%
	98	100%

About two-thirds of the listed consultancies are run by five or fewer people, and almost a half are single-person consultancies. In the 1990s there has been an unprecedented growth in self-employment and chemical engineers are no exception to this entrepreneurial tendency. The first intention of this guide is to provide some assistance and advice to those who, for whatever reason, are considering setting up in business for the first time as consultants, either full-time or part-time. Some of the problems that will usually be faced by new businesses are explored. There is some strictly limited advice and, more importantly, suggestions on where to go for help.

In appendices, the IChemE's policy and bylaws on matters such as professional conduct and advertising have been set out. A draft consultancy services contract is included, together with an example of terms and conditions for work as an expert witness. In addition there is a check list for aspiring consultants to help them see if they really want to enter the field and details of further sources of advice and assistance. In a guide of this sort it is only possible to discuss organizational and administrative topics. Individuals must decide for themselves whether or not their skill is a marketable commodity (in this we advise a strong degree of scepticism and some careful market research before any irrevocable commitments are entered into) and they must also decide whether they are temperamentally suited to the pressure that sooner or later falls on all consultants. The combination of technical, communications and entrepreneurial skills needed by the successful consultant are not so very common. All the skills are needed if a career in consultancy is to succeed.

2. WHAT IS A CONSULTANT?

A consultant is someone employed by a client to provide a special skill or service which the client needs but may not possess. Sometimes the client may have the skills but requires an independent opinion. This is increasingly common in growing areas such as new technology assessment, environmental matters and investment where company policy requires that proposed changes or new investment be preceded by a process of 'due diligence'. This frequently means an independent evaluation of the proposals. A consultant may be required to meet a problem of temporary overload — peak-shaving, one might say — but this is more the domain of the contract engineering group than the consultant. Consultancy is frequently about providing new or extended skills and experience as an extension of the knowledge already available to the client. The consultant is therefore frequently acting in a 'problem solving' capacity.

WHAT ARE THE MAIN REQUIREMENTS?

The first point to be made is that consultancy is not a soft option or easy money in retirement. For the majority of those taking it up for the first time, it requires the special disciplines of working by oneself. Self-motivation is an important factor. Clients do not normally require daily contact with their consultant and you must be prepared to set a routine and adhere to it; it is also important to be able to work on your own. Even in slightly larger practices with several partners, each partner is generally expected to work without constant prompting. If you cannot work on your own, then consider your plans very carefully before embarking on consultancy as a career.

The second point is that a major and important part of consultancy involves communicating, both orally and by writing. The best advice in the world is wasted if the consultant cannot communicate the results of a study to the client clearly, unambiguously and as far as possible in an interesting way. For a summary of good advice read the IChemE booklet *Communication Skills for Engineers and Scientists*. You might even want to recommend it to your clients! For many engineers communication is a problem. For consultants it is the very stuff of life. If you cannot communicate well, consultancy is not for you.

The third point is that a consultancy practice, be it a sole consultant or a multi-partner practice, is first and foremost a business. There are definite legal requirements determining the basics of its operations, although for a one-person operation these are straightforward. There is the overriding requirement that it make sufficient money to pay for its operating costs (stationery, phone bills, heating, lighting, etc) and provide the consultant's income. This will only happen if:

- the basic business idea is sound and properly marketed;
- the business is properly run;
- the consultant is flexible;
- the consultant's skills are such as to command a proper price.

These basic rules sound obvious, yet they may be all too easily forgotten in the enthusiasm of setting out on a new venture, which may be poles apart from an earlier career in, say, a major company. Excitement is one thing. An adequate bank balance is quite another!

Flexibility and the ability to spot and use unexpected opportunities are important characteristics. It is a common experience of many consultants that they find themselves undertaking very different projects from those that they anticipated working on when they first set up their business. It comes as something of a surprise to learn that, although you know very clearly what your skills and experience are, it is far less clear how others can benefit from those skills and that experience.

Another very important factor, in addition to the skills and experience of a newly-started consultant, is the 'network' of contacts which you have established during the acquisition of the skills and experience. In a rapidly changing profession such as chemical engineering, the half-life of the engineer's knowledge is reckoned to be about seven years. A young engineer straight out of university will generally be a better theoretical engineer than the older practical engineer with fifteen years' experience (even with the Continuing Education Programmes which are now a normal feature of the big professional engineering institutions). However, the new graduate lacks two things which are of vital importance to the consultant:

- wide practical experience;
- a network of professional contacts.

The measures required to generate sufficent income include the 'proper marketing' of the consultancy. Nevertheless a surprising number of smaller consultancies find that they require only a low level of advertising;

some never advertise at all, although most will appear in one or two reputable directories. Two major sources of new work are satisfied clients who either provide repeat business or recommend a consultancy, and information on the consultant's personal 'network' which has been established over many years of working in a particular engineering speciality. If you have acquired an international reputation this will also bring work (Trevor Kletz in the field of safety is an example well known to chemical engineers), but for the few rather than for the majority of those who embark on consultancy as a new career.

WHAT TYPE OF CONSULTANCY?

A consultant may practise under a variety of names: consulting engineer, engineering consultant, process consultant, management consultant, even adviser. While there may be differences in the tasks assigned by clients under those names, they are not used in this guide. But it is worth remembering that a consultant, even in a specialized technical area, is likely to be faced with a range of problems from management to 'state-of-the-art' research. All this occurs on top of the need to run the business, generate sales and simultaneously be the office assistant and the managing director — at least in a sole consultancy. The consultant's life may be demanding. It is rarely dull.

WHAT SERVICE ARE YOU GOING TO SELL?

It may seem very clear to you what your skills are and who your target clients are; the check list in Appendix D (page 69) will help you list them. It may well be that you have experience and skills which you do not consider are particularly useful but which some company somewhere will find valuable. And it is almost certainly the case that there are companies around who can use your skills and knowledge but whom you would never recognize as potential clients.

A great deal of consultancy is applied common sense, and flexibility in recognizing a potential opportunity and a potential client can be important in winning new business. Furthermore, as you work in consultancy you will be adding to your store of knowledge and experience and thus adding to your potential market. Flexibility and a willingness to add to your base-knowledge and skills are valuable characteristics for success in consultancy.

PART-TIME CONSULTANCY

The amount of time that a chemical engineer holding a salaried appointment may give to part-time consulting work is a matter for explicit agreement between engineer and employer and the arrangement should be strictly honoured.

In cases where a consultant is a member of the academic staff of a university or college and has permission to use the facilities in that institution, the rate of payment is often determined by the authorities of that institution.

3. HOW TO START

This chapter describes the types of legal entity for a business and provides some advice on the practical aspects of starting a business. Most people considering starting their own consultancy practice will have had some management and business experience; many of the practical aspects of starting a business are essentially common sense. The use of check lists, publications and the advice of others who have experience in setting up their own business will also help.

If you decide that your special skills and experience allow you to enter private practice, you will need to deal with the minor problems associated with self-employment which previously had been the concern of your employer. Initially, therefore, some time must be devoted to planning the routine matters of business and finance such as company registration (if needed), tax matters including VAT, accounts, expense records — and considering their needs with regard to contract law, indemnity insurance, standard terms of agreements and fees, and — not least — Department of Social Security (DSS) payments.

It is sensible to take some advice on these basic requirements before embarking on private practice. A useful first step would be to study literature aimed at helping those wishing to become self-employed; the four main clearing banks all supply literature on self-employment but the advice from the banks depends very much on the knowledge and ability of the local bank manager or the bank's local small business centre. Probably the most important service required by a small business is that of the accountant, and a good accountant is well worth the fees. It is useful to chose an accountant who has experience of working with self-employed people and small businesses. Like all professionals and consultants, the quality of accountants varies considerably, and the best way to find a suitable accountant is by recommendation.

A good source of advice on the practicalities of setting up in practice as a consultant is those who have done it already; the IChemE *List of Consultants* is an ideal source of names, addresses and telephone numbers of those who have tried and succeeded.

A business can take a number of different types of legal form:

- sole trader;

- partnership;
- private limited company (Ltd);
- public limited company (Plc);
- co-operative.

The most common, and it has to be said the simplest, entity for a new start-up consultancy is the sole trader. The next most common form of legal entity in consultancy is the private limited liability company, then the partnership and finally the Plc; no co-operatives are listed in the IChemE's *List of Consultants 1994/5*.

The breakdown of listed UK consultancies by size and type is given in Table 2. For small consultancies of fewer than five people 57% are sole traders, 37% are private limited companies and 7% are partnerships; for one-person consultancies 74% are sole traders and 19% are private limited companies. The preferred legal entity is clearly for the sole trader, followed by the private limited liability company.

TABLE 2
Breakdown of consultancies listed in the IChemE *List of Consultants 1994/5* by type and size

Size	Type and number				Total number in the size range
	Sole traders	Private limited	Public limited	Partnership	
1	32	8		3	43
2	2	8		1	11
3		3			3
4		3			3
5		2			2
6–50	1	16	1	3	21
51–100		7		1	8
>100		7			7

THE SOLE TRADER

Most people entering into private consultancy prefer to practise as a sole trader (Table 2). There is no legal requirement for setting up in business as a 'sole trader'. The term is a legal one used to identify any one-person business or professional practice which is not registered as a limited company or a co-operative. A partnership may also consist of one person and in such cases is very similar to a sole trader. A business run by a sole trader is, in law, identical with the individual and any liabilities acquired by the business are also fully those of the proprietor of the business. Thus, if John Smith sets up a business called *John Smith Associates* or *John Smith and Company* which is a sole trader business (namely, it is not registered as a limited company or co-operative and is not a partnership) then for all legal and practical purposes *John Smith Associates* or *John Smith and Company are in all respects identical with John Smith himself.*

Despite the term 'sole trader' the business is not necessarily associated with trade. This legal term need not deter a consultant from applying the term 'sole practice' when referring to their activities.

The Companies Act 1981 repealed the Registration of Business Names Act 1916. There is therefore no longer any obligation to register the business name under which it is proposed to practise. Unless a practice is carried out under the name of the owner or owners, however, the Business Names Act 1985 requires that there must be prominently displayed at the place of business, the name(s) and address(es) of the owner(s), the business name and the activity of the business. All business stationery must also carry the name of the business and the name of the proprietor. Failure to comply may lead to business contracts being unenforceable. Further information on the requirements should be sought from the relevant Companies Registration Office at:

- *Crown House, Crown Way, Maindy, Cardiff, CF4 3UZ (tel: 01222 388588)* in England and Wales;
- *102 George Street, Edinburgh, EH2 3DJ, (tel: 0131 225 5774)* in Scotland;
- *IDB House, 64 Chichester Street, Belfast, BT1 4JX, (tel: 01232 234488)* in Northern Ireland.

A sole trader has the advantage of being able to raise loans against personal assets to finance the practice. On the other hand it also allows creditors to make claims on personal possessions and not just against those possessions used for business purposes.

A further advantage of sole trader status is that the annual accounts are very simple — nothing more than a listing of income and expenditure. Such a listing can be easily maintained on a computer spreadsheet and is all that is legally required. This means that, if you are diligent, a book-keeper may not be necessary. Your accountant will want to see both the accounts and receipts and any other relevant evidence of expenditure, however, in order to audit the accounts and certify to the Inland Revenue that the accounts are a true statement of the business's financial position. Since keeping accounts is not a skill which interests the average chemical engineering consultant, a book-keeper may well help to avoid problems with the accountant and the VAT office.

PARTNERSHIP

Partnerships are governed by the Partnership Act of 1890 which states that two or more persons carrying on business for profit constitutes a partnership. Twenty is the maximum number of partners for most businesses (but not accountants, solicitors, patent agents and one or two other categories of business). The legal requirements for setting up in business as a partnership are basically the same as those for a sole trader, except for the requirement that a Deed of Partnership must be drawn up; you will need advice from a solicitor to do this.

If you are considering partnership with other people, it will be important to the success of the practice that you first establish how well the partners are likely to work together as a team. Before formally establishing a partnership it is advisable to collaborate on a few assignments with your prospective partners using a prior informal, but written, agreement on how costs and profits are to be shared. In this way you should learn whether or not you are able to work together harmoniously; at the same time you will learn something about each others' strengths and weaknesses. There is, of course, nothing to prevent consultants working together as associates indefinitely on appropriate assignments. In such cases, however — unless accounts are rendered separately, each on the respective consultant's headed paper — there is a risk of being deemed a partnership in law and thus sharing each others' liabilities as explained below. There are also problems of credibility in presenting the consultancy to clients. The 1890 Partnership Act makes all members of a partnership personally liable for all obligations and financial debts incurred by the practice, whether or not the obligation or debt arose from the fault of only one of the partners. All partners are also liable for wrongs done to a third party by a partner in the course of their business, and for claims from negligence, errors or omissions in the professional conduct of the practice by any partner or any employee.

Another type of partnership which should be mentioned is one set up under the 1907 Limited Partnership Act. In this type at least one partner is a general partner with unlimited liability, and one or more partners are limited partners and are liable only to the extent of their financial contribution. Here the limited partners are not permitted to take part in the business affairs of the practice, or they would lose the protection of limited liability. This arrangement has no practical advantage over that of a private limited company.

PRIVATE LIMITED COMPANIES

The IChemE has no rules which would prevent its members from forming a limited company but the Association of Consulting Engineers (Chapter 14 page 45) has such a rule in relation to work on sites in the United Kingdom.The conduct of a practice must, however, conform with the Professional Rules of Conduct set out in the IChemE's Bylaws which are discussed in Chapter 5 (page 23) of this guide.

The main advantages of a limited company are that the claims of creditors are limited to the assets of the company and that no claims can be made against a shareholder's personal assets unless the company has traded fraudulently. In other words the company has a separate identity from that of the shareholders as individuals.

The private company has a statutory minimum of two members. It follows that the protection of limited liability will be lost if the private company is reduced to a solitary member. Table 2 (page 8) shows that eight private limited consultancies in the IChemE's *List of Consultants 1994/5* appear to have only one member. The reason for this is that often in small companies it is common practice for the practising consultant to use a non-consulting family member (for example, a husband or wife) as the other member of the limited company. In this way a private limited company may legitimately have only one practising consultant.

There are a number of important legal requirements to satisfy in setting up a company apart from its registration. Documents to be prepared include:

- the Memorandum;
- Articles of Association;
- Share Transfer Forms;
- Certificates of Incorporation and minute book.

It is relatively easy to set up a private limited liability company without the aid of a solicitor. But it is equally easy, almost as cheap and less time consuming to buy an off-the-shelf company (see the small advertisements in the financial pages of most quality daily newspapers) and then to rename the company with your chosen trading name.

A company limited by guarantee is a variation permitted by company law. In this case one or more directors give a personal guarantee which commits their personal assets with the company funds to the claims of creditors.

Private companies need not be limited companies. The private unlimited liability company is seen by some consultants as having advantages that lie between a partnership and a company, by having shares involved and having a company structure.

The selection of the form of company depends on your preferences, the nature of the risks in the business, the business projections and so on. In contrast to that of a sole trader, the form of accounts for a limited company is laid down by law and accounts must be filed annually at the Companies Registration Office where they are open for public inspection. Also, the closing down of a limited company is more difficult than for a sole trader. So if you anticipate ceasing to trade at some future date there may a disadvantage in trading through a limited company. National Insurance contributions cost more, but there are also more state benefits if you are unable to work for any reason. If, on the other hand, you envisage building up a significant practice then it will be easier to sell the practice as a going concern if it is a limited company rather than a sole trader.

You should seek the best possible advice at a very early stage from a competent accountant and a solicitor, and consider the implications very carefully. Once the form of company has been selected it is a major exercise to change it.

THE CO-OPERATIVE

A co-operative is a business which is jointly owned by its members. Each member has a vote but there may well be appointed managers to conduct the business of the co-operative. A co-operative may be registered under the Industrial and Provident Societies Act 1965 and, if it is, it must have at least seven members. Individual members may have limited liability. A co-operative must be registered with the:

- *Registrar of Friendly Societies, 15 Great Marlborough Street, London W1V 2LL (tel: 0171 437 9992)* in England, Wales and Northern Ireland;
- *Assistant Registrar for Scotland, 58 Frederick Street, Edinburgh EH2 1NB (tel: 0131 226 3224)* in Scotland.

OTHER POINTS TO CONSIDER

It is essential to have adequate financial resources during the initial build-up phase of the practice. This may be possible if you have taken redundancy payments on early retirement. You should also be prepared for a reduction in your standard of living, however temporary it may be. In considering finance remember that unavailability due to accident or illness will almost certainly stop the flow of fees. Many businesses fail either by being under-financed or by not managing their cash flow, particularly in the early years. In addition to an accountant, you can obtain advice from a bank manager or from a number of government-sponsored schemes providing training and information on setting up a small business. Remember, before you approach your bank manager or anyone else for finances, you must have a credible business plan. If in doubt, seek help in preparing it, usually from a friendly accountant.

If you can obtain work for your consultancy practice before it starts up, this is an obvious advantage in overcoming the initial problems of financing. Your most recent employer may provide this, although if you continue to undertake some work for a previous employer on a self-employed basis the Inland Revenue may not accept that you are self-employed. In order to establish that you are self-employed it is necessary to show that you have other paying clients or are actively and realistically seeking additional paying clients. If you cannot do this you will be regarded as still being employed by your previous company. There is also the danger that if you work for a client for an extended period of time, and in that period do not undertake work for any other client, then you may also be regarded as an employee of that client with associated tax and DSS problems.

USE OF TITLES

Members of the IChemE should study Bylaw 9 which is reproduced in Appendix A (page 51). This Bylaw regulates the use of members' titles on business literature, visiting cards, etc.

USE OF SITE BOARDS

IChemE members practising in the construction industry may wish to use site boards displaying the IChemE logo together with their name and titles. These should comply with Bylaw 9 and the dimensions given in Appendix A. If in doubt, ask the IChemE's Membership Department.

4. SETTING UP AN OFFICE AND SIMPLE OFFICE PROCEDURES

It is much more likely that you will visit your clients than that your clients will visit you. Even if your clients do visit you, all that is necessary is a sensible, dedicated and functional room. Clients judge consultants by the quality of their output, rather than the appearance of their office.

Nevertheless, the office is a major resource for your business. You may be a small business but it is essential that the interface with your client is as professional as possible. You may be working from the spare room at home, but when a client telephones you or requires reports from you the response should be, as far as is practicable, of the same quality as that given by a large company.

In order to achieve this you should have a suitable and dedicated office containing all the essential items of equipment to run a business. It is neither sensible nor practicable to use a room at home which has other functions — for example, as a living room in the evenings and an office during the working day. If you are fortunate enough to have a spare room in your home which can be used exclusively as an office, this may be an ideal solution. It does depend, however, on how good you are at personal discipline. Some people find that when working at home they are unable to get down to work; they are distracted by domestic chores and family members who do not appreciate that they need to be left alone to get on with the job of running a business.

If you have a spare room and can be sure of no interference with your working day from domestic influences, and you have the personal discipline to apply yourself to the work of the business whilst still at home, then this is a major advantage since there will be no immediate cost. If you do not have this advantage, it may be possible to rent a room in the offices of an existing company. This has the advantage that many of the necessary facilities will be available. Small and fully-equipped start-up offices are also available for rent; the local press is a good source of advertisements for these. Unfurnished office space is also relatively easy to find. The disadvantage of renting office space is that you must pay in advance and that business rates may have to be paid on office accommodation. If you work from home as a self-employed person business rates are not usually charged — even if you employ one or two additional staff. This can be a major advantage in some parts of the country where local rates make it almost impossible to start up a new business.

You will require the usual office furniture and equipment:

- desk and chair(s);
- filing cabinets and book shelves;
- computer;
- telephones and a fax machine;
- photocopier;
- stationery (notepaper, files, hanging files, folders, index cards, envelopes).

New office furniture is expensive and it is more economical to purchase from companies supplying reconditioned office furniture.

A fax machine is an essential item of equipment and you should not try to run a business without one. Virtually all companies in developed countries use the fax system for communication and it can be cheaper and more reliable than the Royal Mail. It is essential that you make yourself readily accessible to potential clients and that you respond quickly to requests for information and advice. A characteristic of consultancy is that help is often needed rapidly, and any consultant who cannot be contacted quickly and respond rapidly is less useful than one who can.

A dedicated fax telephone line is sensible if you can arrange it. This can be done quickly and at relatively little cost. Although it is possible to get devices which will interpret an incoming telephone call and divert it to the fax machine or telephone, it is generally better in a business — even a small business — to have dedicated telephone and fax lines.

THE TELEPHONE

The telephone is now a much more sophisticated instrument than it was only a decade ago and, although Europeans still do not use the telephone to the extent that Americans do, the range of special facilities available today is invaluable to the small business.

Most business communications are via the telephone network. In addition to simple person-to-person voice communication, the telephone network now allows a range of additional forms of communication. Examples of these include — but are not limited to:

- document transmission via the fax machine or PC (personal computer);
- transmission and receipt of documents and data via electronic mail systems;

- voice-mail systems;
- access to databases;
- access to information networks (for example, the Internet);
- telephone conferences;
- portable telephones.

All can add greatly to the efficiency and success of a small business and, although some of them may seem complicated, many of these communications and information technology (IT) systems will quickly pay back the time spent in learning how to use them.

In the UK British Telecom (BT) offers a number of useful services called the Star Services. A modern telephone will normally include some of them, such as *repeat last call* and *code calling*. Others are invaluable; the two most useful Star Service facilities are *call waiting* and *call diversion*.

Call waiting is a method of allowing you to answer an incoming call if you are already on the phone when you are called. Whenever anyone tries to get through when your number is busy, they receive a message asking the caller to hold on and saying that the system is attempting to contact you. At the same time the system plays a series of bleeps which only you can hear. You can put the current call on hold and answer the incoming call and take a name and number in order to return the call. Then you can return to the original call. This ensures that you can use the telephone even when waiting for an urgent call. It also ensures that you do not miss out on a potential client who is making a series of speculative enquiry calls.

Call diversion is essential for all small businesses. It allows you to divert incoming calls from your telephone to any other telephone in the world. If you are away from your office for a period of time you can divert your calls to a portable telephone or to another telephone where there is someone who can take a message for you. The transfer is made seamlessly and the caller is unaware that a call has been diverted. This facility is necessary if you use an answering service.

PORTABLE TELEPHONES

There are two types of portable telephone; both of which can be valuable to a small business:

- domestic extensions (or cordless telephone);
- network portables.

The domestic extension is simply a cordless extension of the desk top phone, but it is invaluable when you have to move around whilst talking to a client. If you are alone and happen to be answering a call of nature when the phone rings, a cordless phone can save considerable inconvenience. It can also avoid the impression that your office is unattended (a remarkable number of people will only allow the phone to make three or four double rings before they assume that it will not be answered and hang up). It sometimes happens that you are on the phone to an important client when a delivery of goods is made and you have to answer the door. A cordless phone makes this possible without having to break contact with the client.

A number of different network portables are currently available. All have a voice-mail facility so that if the number is unavailable when called (for example, if you are in a train in a tunnel or in a building with a substantial metal structure) the incoming call can be automatically diverted to a voice mailbox where a message can be left; the presence of the message is indicated on the handset when the portable phone is next in contact with the network.

The cost of using different portable phones varies considerably. Careful analysis of your potential use and a simple cost-benefit analysis is advisable before buying one. A major benefit is that, in combination with the BT *call diversion* facility, it is possible to remain contactable by clients who call your office when you are out.

ALTERNATIVE TELEPHONE NETWORKS

In addition to BT there are a number of alternative systems. One of the best known alternatives at the time of writing is Mercury Communications.

A major advantage of Mercury over BT is that the Mercury system allows you to allocate cost codes and receive invoices with all the calls for each code on a separate sheet. This has clear advantages in allocating costs to clients.

Another inexpensive system for international telephone calls available in the London area is Swiftcall. For the cost of a local call you are connected to an exchange in New York whence you can dial anywhere in the world at cheaper American rates. This system allows you, for example, to use the toll-free 800 numbers in the US so it is often possible to call an American company for the cost of only a local call.

For overseas calls BT, Mercury, Interglobe Telecommunications (UK) Plc and the American company AT&T, amongst others, offer call cards. Most of these cards make it possible to call from and to anywhere in the world without using cash, and to be billed monthly. Call cards work by calling a toll-free number in the country from which the call is being made and then asking for, or dialling, the required number. Be careful in hotels, though. Many now program their telephone exchange systems to recognize these toll-free numbers and impose a high charge for their use on customers' bills. Another system which avoids the high costs of many hotel telephone systems is Kallback Direct of Seattle in the USA. With this system you call the system number and hang up; the system then calls you back and connects you with an exchange so that you can then dial the number you require. The hotel exchange regards this an incoming call and does not charge for it.

This area is in rapid change; you will need to get fully up to date with all options before making a decision.

THE COMPUTER

A computer is an essential item of office equipment for anyone undertaking a serious consultancy practice. It enables reports to be produced in a professional manner, facilitates complex calculations, simplifies accounting, can be used for searching databases from the office and permits documents and information to be exchanged in non-paper formats (as floppy discs or e-mail). A good laser printer is also advisable because the quality of the output helps to make documents look professional.

The choice of computer is, to an extent, a personal matter. There are two main types of office computer: the PC and the Macintosh. The PC was developed by IBM but was copied and then developed beyond recognition by a host of companies producing variations on the original. Many of these so-called 'clones' are excellent value and the best advice is to talk to someone who has bought a few computers and to consult computer magazines. Beware, though. Magazine articles are written by people for whom the computer and its architecture is the end-purpose, and who often have no experience of using a computer as a business tool. Also, they are often written in an opaque jargon which is frequently incomprehensible unless you, too, are a computer enthusiast.

The Macintosh computers, developed by Apple, are said to be much easier to use but until recently they were much more expensive than PCs and did not represent good value for money unless you had some specific uses for which the Macintosh computer was superior. For example, Apple was the first to introduce integrated software which enabled data produced by, say, a mathematical program to be incorporated into a document produced by another piece of software. Microsoft, the company which produces most of the operating system software for PCs, also uses this concept. You can now buy integrated software packages for PCs incorporating four or five programs including a word processor, a spreadsheet, a database, a desktop publishing package and, possibly, electronic mailing. This may be helpful but is not essential.

The minimum software required is a word processing package and a spreadsheet package. A desktop publishing package can be very helpful in improving the presentation of documents and presentations to clients. A database package will rapidly become a useful item of software as you collect data and information in your particular speciality. If you become involved in IT and information database searching, an electronic mailing package and software which enables you to search databases will become essential.

There are also a number of excellent accounting packages available at modest cost. They help a good deal in keeping the books, including VAT monitoring and returns. These packages simplify audit work for your accountant (and so reduce accountancy fees!). They also have the benefit of being easy to use.

As with most items of equipment, the best way of deciding which of the many types of computer and software to select is to talk to people whose judgement you trust, seek advice from at least two or three suppliers and try out one or two systems if possible.

THE PRINTER

There are three main types of printer currently available: dot matrix, laser and inkjet printers. The quality of the dot matrix printer makes it unsuitable for professional reports, and the choice lies between laser and inkjet printers. The latter are cheaper and quiet; the former are more robust and produce a better quality product. Both are normally limited to paper sizes of A4 and below.

Both can produce drawings, but if you are likely to produce many drawings it may be worthwhile considering a plotter as well as a printer. If you need to produce drawings larger than A4, a plotter is essential, together with the appropriate software package for producing drawings.

Laser printers require large cartridges of powder whereas inkjet printers usc small cartridges of ink. The replacement costs for these items can vary substantially and it is worth shopping around.

STATIONERY

Stationery can be surprisingly expensive, and it is worthwhile to seek a reliable but lower-cost supplier. There are a number of national suppliers of bulk stationery who will deliver to the door. If you can plan your replacement needs, this is a time- and cost-saving way of purchasing stationery and other office consumables.

Letterheads are expensive to have printed and there is a risk that your address or phone number may change before all the printed stock is used up. For instance, over recent years BT has changed many telephone numbers; alternatively, if your business grows, you may need new premises. With a good word processing package it is possible to set up the word processor so that it automatically prints the letterhead in a professional way on blank paper — or on paper which carries just your logo and nothing else.

ROUTINE OFFICE MANAGEMENT

When you run a single-person business you have to undertake most of those essential tasks which, in a larger company, are looked after by invisible specialists. It is not only secretarial and routine filing activities and possibly bookkeeping which you will be undertaking. There is also invoicing, placing orders and keeping records of payments and invoices. If a record of these transactions is not maintained then chaos will result. You may end up paying bills twice, being threatened with court action for non-payment of bills or losing money because clients are slow in paying their bills (and, regrettably, in the UK most clients *are* slow in settling their bills).

You should have a simple system of keeping a record of every order placed in which you note:

- the order number;
- the item ordered;
- the supplier of the ordered item or service;
- the quoted cost;
- the date of receipt of the order;
- the date of payment.

With regular suppliers (such as stationery, telecommunications services or answering service) it can be helpful to have a separate listing of invoices received and when they are paid. All invoices must be kept for the accountant and it can be a very useful practice to write on each invoice the date of payment and the cheque number. This will help both the VAT inspector (who calls at irregular intervals — probably once every year or two) and the accountant by providing an audit trail.

5. PROFESSIONAL CONDUCT

Under the IChemE's Bylaws a member 'shall exercise his professional skill and judgement to the best of his ability and discharge his professional responsibilities with integrity'. This implies that members should not set up a consulting practice in an engineering or related field unless their competence in that field is based on broad experience. Clients are more likely to feel confident in a consultant's ability if that consultant is a corporate member of the IChemE. It is also implied that assignments in areas outside a member's professional competence should not be accepted.

Members in private practice as consultants will probably need a greater awareness of their professional obligations in their relationships with clients than is needed by the majority of other members engaged as employees in industry. You should study the IChemE's regulations governing the conduct of members carefully, as set out in Bylaws 12 and 13 and the associated ten Rules of Conduct. For this purpose these regulations are reproduced in full in Appendix A (page 51).

Although these are the only formal rules of professional conduct set by the IChemE they have a very broad scope indeed, and members are reminded that should they be accused of a breach of rules, it could lead to expulsion or other disciplinary action through the IChemE's Disciplinary Board.

Bylaws 12(ii)(a) and all ten of the Rules of Conduct apply equally to both corporate and non-corporate members and state (*inter alia*) that a member 'shall so order his conduct as to uphold the dignity and reputation of the Institution, and shall observe the provisions of the Charter and these Bylaws'. Corporate members have additional obligations under Bylaw 12 (ii)(b) which includes 'to safeguard the public interest in matters of safety and otherwise'.

Note that the special obligations towards an employer contained in the Rules of Conduct apply equally to professional relations with a client.

Both the Association of Consulting Engineers and the Association of Consulting Scientists (Chapter 14 page 45) publish detailed Codes of Practice to which members seeking advice may refer. Whether they are relevant to the practice of consultancy in the process industries is a matter for individual judgement.

6. ADVERTISING BY CONSULTANTS

There are two main reasons for advertising:

(i) to win business;

(ii) to maintain an image or awareness of the business.

Some clients feel more comfortable with a business which is familiar, even if that familiarity arises out of nothing more than seeing an advertisement regularly in, say, *The Chemical Engineer.*

Advertising can be expensive and it advisable to make sure that you have a clear objective before you start advertising and that you check the success or failure in achieving that objective.

CONTROLS ON ADVERTISING

The IChemE, unlike the Association of Consulting Engineers, has no specific rule banning advertising by consultants. Members must be guided by the IChemE's Regulations governing their conduct (Appendix A page 51). Bylaw 12(ii)(a) and rule 5a are particularly relevant in this context. It will be evident that any public or trade advertising must be discreet and generally confined to a description, free from embellishment, of the specialist engineering services available. Reference to specific projects on which a member either is, or has been, engaged should be avoided. However, it is considered acceptable to describe, with the agreement of the clients concerned, past assignments on projects as part of a consultant's references in a brochure describing the practice for the benefit of potential clients seeking professional services. Such a brochure should also detail the services offered and the precise areas of expertise available. This is an essential document for a consultancy practice and demands a good deal of attention. Business cards should show your professional qualifications.

Advertising will be considered discreditable to the profession of engineering for any of the following reasons and may lead to disciplinary action being taken against the member concerned, where it:

- contains any false, exaggerated or misleading statement or implication;
- contains invidious comparisons, or is derogatory towards other consulting

engineers or professionals;

- is likely to cause significant offence among other members;
- uses self-laudatory language.

A member who is the principal or one of the principals of a practice that is not constituted as a limited company, or who is a director of a company so constituted, is held to be responsible for the advertising of a consulting practice.

Members who need advice on whether a particular type of advertising is likely to be acceptable to the IChemE should make their enquiries through the IChemE's Membership Department. For the practice to prosper it is necessary to let potential clients know of your capabilities. This is a time consuming and expensive process but it must be done. Work does not walk through the door!

INFORMAL ADVERTISING

Working on the technical committees of, for example, the Department of Trade and Industry, British Standards Institution, IChemE or universities may provide valuable business contacts.

One means of promoting a professional practice can be through close involvement in the affairs of the IChemE. Engineers are likely to become known if they take a leading part in branch and headquarters meetings or conferences and either prepare and deliver authoritative papers or take an active part in the discussions. Whether or not a conscious effort is made to become known in this way, it is still necessary for consulting engineers to keep abreast of all new developments in their field of practice and this should include active participation in appropriate technical meetings.

DIRECTORIES

Other sources of information sometimes used by potential clients are trade and other directories. There are a great many directories and the value of some of them is doubtful. If you are just one of a hundred or so names in a directory your chance of being selected is remote. Some consultants include themselves simply to ensure that if a client looks in the directory there is 'feel-good' or 'reassurance' factor. Many directories charge nothing, and the only cost is in preparing the data for the entry. As a rule of thumb, if a directory charges for inclusion you should have doubts about its usefulness and validity, and consider the cost-benefit very carefully before paying. Directory salesmen can be very persuasive; there are a great many directories and it is easy to end up paying substantial sums for little or no benefit.

The most obvious example of a useful directory is the IChemE *List of Consultants* which is produced every other year. There is a fee for inclusion, but there is good evidence that the *List* is well-used. It is also a useful source for obtaining additional services when you have a project which includes topics on which you are not expert.

Some of the better directories are selective about whom they include, and require evidence of successful projects before they will include a consultant. Similarly, the IChemE will normally only include consultancies which use or employ chartered chemical engineers.

ADVERTISING IN JOURNALS

Most trade journals carry pages of consultants and if the area of your expertise is represented by trade journals this may be a useful place to put your advertisements. However, the cost of monthly advertisement can soon mount up and it advisable to monitor the degree of response to an advertisement and to see if the work won by it pays the costs of advertising.

Remember that once you have placed an advertisement in a journal it will be noted by the advertising managers of other journals who will then start to pressure you into advertising in their journals as well. There will be times when you must be firm.

One final point to note is that a number of the apparently successful consultancies listed in the IChemE *List of Consultants* do not advertise anywhere else. Advertising can be expensive, especially for a one-man business, and it is not necessarily essential for winning new business.

7. CONDITIONS OF ENGAGEMENTS

It is vital to any engineer starting up a consultancy practice to have a sound understanding of the legal relationship with their clients. Large firms engaging consultants frequently have their own forms of agreement which are sacrosanct. In these cases it is important to study all the conditions, particularly those in small print which may concern penalties for non-delivery of drawings or reports, or responsibility for events which take place after the project is completed and which are outside the consultant's control. Measures to meet such penalizing clauses should be assessed most carefully so as to avoid potentially serious financial penalties.

It is always sound practice for consulting engineers to enter into written agreements with their clients which identify the main conditions of engagement. This reduces the risk of subsequent disagreement over the services provided and the payment of professional fees.

Some consultants are able to agree on many assignments by a simple exchange of letters saying what is to be done, by when, and for how much. In drafting agreements for complex assignments, however, the following check list is useful:

- identification of the contracting parties;
- description of the assignment to be undertaken — worded with care;
- scope of services to be provided — also needs careful wording;
- style and language of report required (and number of copies);
- ownership of documents and inventions;
- premature termination of the contract;
- payment for services — rate, progress payments, etc;
- disbursements — allowable expenses, sub-consultant fees, etc;
- settlement of disputes — arbitration;
- signatures.

Consultants should take care not to place orders in their own name for the supply or manufacture of items needed by the client in the course of a project. Orders should be placed only in the client's own name and with the client's authority if liability for some unforeseen failure in supplies is to be avoided. An exception to this may arise if the consultant is working as an Expert Witness (Chapter 12 page 41) in a legal case and requires additional specialist work to be undertaken in support of the Expert Report. Any order placed under these circumstances should be placed by the consultant; this preserves and demonstrates the independence of the consultant (as an Expert Witness) from the client who is paying the consultant, and shows that full control is exerted by the consultant over all the information used in preparing the Expert Report.

If you are unfamiliar with suitable agreements for consultancy work, seek advice. Failure to do so can prove very expensive in the long run.

Whatever is agreed, set it down in writing before starting work and have it signed by both sides. Good will is not enough in contracts.

8. FEES

Fees are a matter for negotiation between the consultant and the client. They take account of the nature of the work and the degree of special expertise that the consultant has to offer and the 'going rate' in the market for that type of work. Although fees are often charged hourly or daily, they may also be charged as a lump sum against an agreed scope of work. It may also be possible on occasion to negotiate retainer fees. The IChemE does not publish a scale of fees for consultants. Fees are influenced by three considerations:

- the marketability of the particular field in which the consultant practises;
- the level of expertise the consultant is able to offer;
- what the client is willing to pay for the service.

As a guide to arriving at a fee level in the early days of a sole practice, there are four main elements to be looked at:

- the taxable income which the consultant expects to earn in order to maintain a reasonable standard of living;
- the number of working hours on client's business available during a year;
- the allowance for the overhead expenses of the practice;
- the need to price services according to the market.

Underpricing can be as destructive as overpricing.

TAXABLE INCOME

On this important point you must exercise your own judgement based on your knowledge of the current salaries of others in your field of practice, on your own experience and on your professional ability. Initially you will, no doubt, consider the salary earned in the recent past as a yardstick.

ANNUAL WORKING HOURS AVAILABLE

As a guide, a consultant in sole practice should budget to be able to provide about 2000 working hours per year, of which between 1000 and 1200 hours per year will be chargeable time. The rest of your working hours will be taken up in running the business, sales, developing your own professional skills, attending conferences and generally maintaining your profile in the profession. It is only chargeable time that can be considered when calculating billing rates.

OVERHEADS

It is most important to make an accurate assessment of annual overhead expenses as this will also need the agreement of the tax inspector. It is essential to engage the help of an accountant in making a list of items allowable against the expense of running the practice. These include the routine items such as office rent, fire and theft insurance, heating, lighting, cleaning, telephone, stationery and secretarial services. Account must also be taken of the cost of maintaining a reference library for the consultancy with books and journals, attending conferences, professional subscriptions, and business equipment such as computers and portable telephones. There is also the question of travel — including, for example, a car. Typically overheads are at about the same level as the basic salary bill. Any profit the business is to make comes after that.

CALCULATING A STANDARD FEE

Having established the revenue needed, calculate the hourly or daily fee simply by dividing the income by between 1000 and 1200 hours and adding any percentage budgeted for profits. A billable working day is typically between 7.5 and 8 hours.

EXAMPLE:

If T = salary income (taxable income), H = percentage overhead charge and P = percentage profit, then assuming T = £30,000 per year, a figure which varies between consultants and with the cost of living, H = 100% and P = 25%, then, for 1000 hours per year, the rate becomes £67.50 per hour or £506–£540 per day.

For a single-person consultancy the concept of profit is relevant, but the main concern is to achieve a specified income. Any money for the writing off of capital equipment should be included in the overheads figure.

Thus, if an income of £30,000 per year is required and the anticipated earning year is 1000 hours and the overheads are 100%, then the hourly rate would be £60 per hour (daily rate of £450–£480) or, for 1200 hours per year, £50 per hour (daily rate of £375–£400).

Travel, hotel and other out-of-pocket expenses as well as disbursements are usually additional to the negotiated fee. They are normally shown separately on the account, although a fixed fee job could well include all travel and accommodation charges. Estimate these charges carefully, particularly if air travel is involved. Clients should be made aware of this arrangement during the fee negotiations. On some projects expenses can substantially exceed 10% of total billings.

It is important to charge at rates which will generate profit because profit is the only internal source of funds available to develop the business for the future. Repayment of external loans is a charge against profit and it is also desirable to build up some reserves to guard against inevitable periods of lower activity. Overall preparation of a realistic budget is a key activity. In the early days of a practice be careful not to be too ambitious over expenses.

If the consultancy is registered for VAT the sums given all exclude VAT. Expenses invoices should thus be invoiced nett of VAT (that is, excluding VAT which will be claimed back by the consultancy).

OVERSEAS WORK

For work outside the United Kingdom, the fees can be higher than those calculated in the previous section. You should take account of the local market conditions and the cost of living in the country concerned. You should be aware of the risks involved in rapidly changing rates of exchange and that the level of out-of-pocket expenses may become a significant proportion of the reimbursement. Middle-East clients frequently demand performance bonds to be deposited through a bank. It is always possible to take out insurance cover in the event of a performance bond being called in by the client, but the premium will need to be taken into account when negotiating the fee structure for the project.

Assistance is available from the Export Credit Guarantees Department (ECGD) (part of the UK Department of Trade and Industry at No.2 Exchange Tower, Harbour Exchange Square, London E14 9GS, tel: 0171 512 7000, fax: 0171 512 7649). ECGD will insure exporters against the risk of not being paid by an overseas buyer or supply a guarantee to an exporter's bank against which security the bank may provide finance at favourable interest rates.

FEE PAYMENT

During the negotiation of fees for long-term projects, consultants are advised to arrange to invoice their fees monthly and ask for payment within, say, 30 calendar days. In some cases the payment of fees may be linked to project stages or reporting stages. In some parts of the world insistence on a Confirmed Irrevocable Letter of Credit may be the only practical way of ensuring payment.

EXPLORATORY VISITS

You may feel that it is in your interest not to charge a fee for a preliminary exploratory discussion with a prospective client. But take care before committing too much of the practice's money to expensive sales exercises.

FEES FOR COURT APPEARANCES

Appearances in Court as an Expert Witness should be charged at the same rate as for normal assignments (bearing in mind the degree of specialist expertise involved), but it is suggested that the minimum fee should be equivalent to five hours' work plus any additional time necessarily devoted to case study if not already accounted for elsewhere. In cases of briefing by a solicitor it is sometimes specified that the solicitor will pay only after receipt of fees. The consultant should be cautious about accepting such terms since many solicitors invoice infrequently — some do it only every six months — and the consultant may find that it takes six or seven months before fees are paid if such terms are accepted. Law firms are usually wealthier, and have better cash-flow, than sole practice consultancies and it is advisable to insist on nett monthly payment. The Law Society's rules require solicitors to pay their sub-contractors whether or not the solicitor has been paid. An example of terms and conditions for work as an Expert Witness is reproduced in Appendix C (page 65).

9. INSURANCE

As self-employed persons, consultants in private practice are responsible for their own insurance cover. In particular it is a statutory requirement for the self-employed to contribute to the National Insurance Scheme. The simplest way to avoid problems and ensure regular payment of the correct sum is to raise a direct debit order in favour of the Department of Social Security (DSS). Details of payments and direct debits facilities can be obtained from the local DSS office.

Professional Indemnity Insurance is a voluntary insurance protecting the insured against claims arising from negligence. A consultant is responsible for the results of the advice or service given to his clients. If personal injury or damage to property can be shown to be a direct result of any negligent act, error or omission in the professional conduct of your practice then you are liable to be sued for damages by the injured party. Claims will be directly related to the loss suffered by the claimant and may therefore be much greater than the consultancy fee. There have been cases of consultants being sued for damages although they displayed no negligence or omission, and of insurance companies choosing to settle the claim out of court rather than enter into a complex legal wrangle.

A recent court case has held that consultants are responsible for their professional acts even into retirement up to the time of their death. Indemnity insurance should therefore provide whole life cover.

Professional Indemnity Insurance can be sought either by project or to provide annual cover for the practice. It is for you to seek the type of cover that is the most economical to suit your special needs and risks. A feature of indemnity policies which is particularly valuable in protecting the reputation of a practice is that the consultant shall not be required to contest any legal proceedings unless a Queen's Counsel advises that contesting the legal proceedings is likely to be successful. Other important features of these policies which may be negotiated are:

- scope of cover provided: damages, claimant's costs and expenses, cost of defending or negotiating a claim;
- consequential liability of the consultant for damages arising from the negligence of sub-consultants;

- the limit of indemnity;
- the excess — the sum for which the consultant accepts responsibility;
- cover for work outside the United Kingdom;
- partners' previous business — liability incurred by partners in the business or practice for claims arising out of a previous business or practice in the same profession;
- former partners — liability incurred by former partners for claims arising out of the business or practice;
- documents — loss of or damage to documents in the policy holders' custody or in transit;
- libel and slander;
- dishonesty of employees — liability for breach of professional duty arising from dishonesty of employees.

It is up to you to negotiate special additional requirements or exclusions to suit your particular needs. However, professional negligence policies are expensive. You may wish to consider limiting your liability by contract with your clients. This is a common practice in the process industries. The issue of liability also impinges on the form of company (for example, limited company) set up initially. This is another area where advice should be sought.

Other insurance policies which will probably be needed are:

- employers' liability insurance to cover the statutory obligation to insure employees against death, injury or disease;
- public liability insurance — to cover accidental injury to persons other than employees;
- fire and theft insurance — to cover fire, theft or damage to the premises and effects of the practice.

Competitive quotations should be sought from reputable insurance brokers; take care not to base a choice only on the lowest premium offered. When faced with a claim the consultant does not want arguments about its validity arising from the small print in the contract. The Professional Engineers Insurance Brokers (PEIB) Ltd is approved by the Institution for any insurance cover likely to be required by members. The PEIB professional indemnity policy covers the first five features listed above as standard terms in the policy.

A list of brokers who provide professional indemnity insurance for consultant engineers is maintained by the IChemE's Technical Department, and is available upon request.

10. ACCOUNTANCY AND TAX

A consultant will need to adopt a suitable accounting system, however simple, right from the start of private practice. Every expense and VAT payment should be recorded at the time when it occurs, to avoid them being overlooked when cash books and accounts are subsequently made up.

It is necessary to register for VAT purposes where the annual turnover of the practice exceeds a specified amount. The latest threshold figure above which VAT registration is required can be obtained from an accountant. All transactions involving VAT need to be carefully recorded so that allowances can be made for the VAT paid out when calculating VAT due for payment.

It is also permissible to register for VAT where the turnover of the practice is below the threshold figure. The main advantage in being registered for VAT is that all VAT paid can be reclaimed (usually by being off-set against income received). Although it is also necessary to charge VAT on invoices submitted by the consultancy, the preparation of VAT returns is simple and is not time consuming provided proper accounts have been kept. The nett effect of registering for VAT when the turnover is below the threshold level is to reduce the cost to the client, or to increase by a small amount the consultant's income.

If the practice has clients in other countries in the European Union the rules for VAT charging and reclaiming become more complicated and advice should be sought from the local VAT office. VAT is not chargeable on invoices submitted to clients in countries outside the EU, and in these circumstances there will be a small but significant saving in the expenses charged by a consultant registered for VAT compared with a consultant who is not registered.

There are a number of published guides to the VAT system. An accountant or the VAT authorities can help you further on this subject.

The VAT system is administered by HM Customs and Excise and not the Inland Revenue, and generally officers are very helpful with small businesses, especially when they realize that you have problems and are not dishonest. If you are registered for VAT you should anticipate having your accounts and VAT returns inspected on site by a VAT inspector once every two or three years.

11. INVENTIONS AND COPYRIGHT

The client may ask to be vested with the full benefits of any process, invention, discovery or improvement arising directly from the consultancy work done and relating to the subject of such work. It is recommended that provision is made in advance in the conditions of engagement to cover the ownership of any inventions or discoveries likely to arise and to define the areas of direct interest to the client.

In a partnership, each partner is liable to share any royalties or payments arising from inventions with the other partners in the absence of a written agreement to the contrary.

The ownership of copyright of designs and drawings produced by you should normally be vested in you. This can be an important factor in cases where the client defaults on payment for the consultancy work.

12. THE ROLE OF THE EXPERT WITNESS

Professional engineers may be invited to give expert testimony in a civil action in a Court of Law by one of two parties to a dispute, or in criminal proceedings. In doing so engineers are given the privilege of having their professional opinion taken as evidence. Opinions are not normally permitted in the Courts except in the case of expert testimony. Therefore the privilege granted to the Expert is not applicable to other witnesses who may only testify to facts as they recall them.

The duty of the Expert Witness is to assist the Court (that is, the judge) to reach an equitable verdict. The Expert Witness's role is not to be an advocate for whoever pays a fee. Expert Witnesses are bound by the needs of professional integrity and the rules of the Court to present unbiased views of the facts as they see them in the light of their expert knowledge even if these views are not in the best interests of their client. Experts must, therefore, be careful not to express opinions on matters which lie outside their expertise and special skill. The Expert's testimony must be to the truth as the Expert believes it to be. Experts may advocate any views as strongly as they like on matters within their area of experience but it must be their own view and not the views that the client would like to be put forward.

The Judge is experienced and skilled in the understanding of problems and in the assessment of evidence, but on the technical topic covered in expert testimony the Judge relies on the witness for skilled guidance. Some care should therefore be taken to prepare the evidence to avoid plunging into the technicalities of the case with the likelihood of leading the Court into a hopeless fog. It is good practice to start the evidence with simple guiding principles and then lead into the technical matters relevant to the case.

You should enter the witness box with the right attitude of mind. Bearing in mind all the advice given in this section, remember not to be dogmatic but to be as simple, as brief and as interesting as possible and exhibit courtesy, integrity and reasonableness. If there is a difficulty in the logic of your arguments which you cannot resolve, face it squarely and bring it to the attention of the Court. If the opposing Counsel exposes your difficulty under cross-examination it can throw doubt on the sound opinions which you have honestly advanced. By stating your own difficulty, the Court is likely to make less of it in the context of your whole evidence than would be the case if it were only brought out by the opposing Counsel.

If the opposition produces an Expert Witness with a different opinion on the case from that given in your own evidence, remember that the Court will only be concerned with the reasonableness of the arguments advanced in favour of the opinions expressed and will be unimpressed by any dogmatic stance adopted.

Do not be trapped, under cross examination, into giving simple unqualified answers if they would give rise to the wrong impression. As an Expert Witness you are entitled to say that a simple answer would mislead the Court and to explain why. You would then be in a position to amplify or qualify the answer. As an Expert Witness your attitude of mind and your professional integrity are of utmost importance. If you are called upon to give expert testimony you are advised to read a paper entitled 'The Giving of Evidence' which was delivered to the Institution of Civil Engineers (ICE) by the Right Hon Lord Macmillan of Aberfeldy (available from the ICE, tel: 0171 222 7722, fax: 0171 222 7500). It is worth noting that the clarity and presentation of the consultant's written report frequently leads to cases being settled out of Court. It is therefore worth extra effort to achieve this end.

A professional association, The British Academy of Experts, exists for the Expert Witness. If you are invited to become an Expert Witness you can obtain further information on the subject from the Academy at 116–118 Chancery Lane, London WC2A 1PP, tel: 0171 637 0333, fax: 0171 637 1893.

13. ARBITRATION

Chapter 7 of this guide, 'Conditions of Engagement', makes reference to arbitration in the settlement of disputes. In arbitration agreements the President of the IChemE may be nominated to appoint an arbitrator to effect settlement of a dispute. A fee is payable to the IChemE to cover administrative expenses arising from such an appointment.

The President will normally only appoint as arbitrators Corporate members of the IChemE, on the basis of both their knowledge of the particular issues involved and of arbitration in general.

Assistance and training in arbitration matters can be obtained through the Chartered Institute of Arbitrators at 24 Angel Gate, City Road, London EC1V 2RS, tel: 0171 837 4483, fax: 0171 837 4185. In addition IChemE has published a *Guide for Arbitrators*.

14. CONSULTANTS' ORGANIZATIONS

Association of Consulting Engineers
Alliance House
12 Caxton Street
Westminster
London SW1H 0QL
tel: 0171 222 6557
fax: 0171 222 0750
Consultants are not permitted to join the Association of Consulting Engineers (ACE) until they have been in consulting practice for a number of years. The Association publishes several useful documents which are available to non-members, however, and a list of these may be had from them. For consultants in the construction industry the ACE publishes *Conditions of Engagement* and *Scales of Fees.*

Association of Consulting Scientists
Westgate House
39/41 Romsey Road
Winchester
Hants SO22 5BE
tel and fax: 01962 869128
You will need to investigate for yourself the value of joining the Association of Consulting Scientists (ACS). Some consulting engineers whose field of practice involves scientific laboratory investigations find membership of the ACS beneficial.

British Consultants' Bureau
1 Westminster Palace Gardens
1–7 Artillery Row
London SW1P 1RJ
tel: 0171 222 3651
fax: 0171 222 3664

The British Consultants' Bureau is generally of most use to those consultants engaged in civil or heavy engineering and can be a valuable platform for them from which to seek overseas work. It facilitates the exchange of ideas and uncovers work opportunities overseas. You should investigate for yourselves the value of joining this Bureau.

The Design Council
28 Haymarket
London SW1Y 4SU
tel: 0171 839 8000
fax: 0171 925 2130
The Design Council registers suitable consultants for a registration fee.

Institute of Environmental Assessment
Gregory Croft House
Fen Road
East Kirkby
Lincs PE23 4DB
tel: 01790 763613
fax: 01790 763630
This organization (IEA) was founded to improve standards in environmental assessment and consultancy. Membership is drawn from a wide variety of companies and individuals, and it provides a list of environmental consultants. The IEA also operates a quality standard for environmental consultants through registered membership.

Union Internationale des Laboratoires Independents
David H. Stranger, Secretary General
c/o The Laboratories
Fortune Lane
Elstree
Herts WD6 3HQ
tel: 0181 207 3191
fax: 0181 207 5878
The Union is an association of private (that is, non-governmental) laboratories and independent consultants. The area of activity of members is principally in analysis, testing and laboratory investigations but it covers chemistry and some aspects of chemical engineering.

Most consultants' organizations are currently organized mainly for the more traditional industrial areas. They seem to have difficulty in adjusting to the particular situation and organization of the process industries.

15. CONCLUSION

It is impossible to cover all aspects that could arise in a consultancy practice in a booklet of this length. Much of what is necessary to become a successful consultant, however, is based on the experience gained during a potential consultant's professional life, aided by common sense. It is quite possible to start a consultancy with no more advice than is in this guide, and to succeed. A discussion with consultants who have already gone through the experience of starting up a consultancy business may provide additional information — and many such people can be found in the IChemE's *List of Consultants*.

Finally, readers seeking further information or advice may contact the IChemE's Technical Department.

APPENDIX A — EXTRACTS FROM IChemE BYLAWS

In the following quotations from IChemE Bylaws and Rules of Professional Conduct, the assumption made that members are male rather than female is for convenience only.

BYLAW 9: ABBREVIATED TITLES OR DESCRIPTIONS

i. Each Honorary Fellow being either a Corporate or Non-Corporate Member may use the title of this class or the appropriate abbreviation thereof after his name as shown hereunder, namely:

Honorary Fellow — Hon FIChemE

ii. Each Corporate Member may use the title of the class to which he belongs or the appropriate abbreviation thereof after his name as shown hereunder, namely:

Fellow — FIChemE

Member — MIChemE

iii. Each Corporate Member may describe himself as a Chartered Chemical Engineer.

iv. Each Associate Member may use the title of the class, provided that he is registered as a Technician Engineer under the provisions of the Royal Charter and Bylaws of The Engineering Council, the abbreviation after his name as shown hereunder, namely:

Associate Member — TEng AMI ChemE

v. No other title or abbreviation shall be used to describe membership of the Institution.

vi. Where a Corporate Member uses the description Chartered Chemical

Engineer and also the abbreviated title of his class he shall place the description after the abbreviated title.

vii. Any Corporate Member or Non-Corporate Member of the Institution who is entitled to be registered under the provisions of the Royal Charter and Bylaws of The Engineering Council and wishes to describe himself as such and at the same time describe himself as a member of the Institution may so long as he shall remain a member of the Institution use such style or title of the same Council or the approved abbreviation thereof followed by the appropriate title or abbreviation (if any) authorised under these Bylaws.

viii. A member shall not use or permit to be used any of the said titles or descriptions in letters larger or bolder than those used in the name of the member which they follow.

ix. A member practising (a) under the title of or as an officer or employee of a limited company or (b) in partnership with any person who is not a Corporate Member shall not use or authorise to be used as descriptive of any such company or partnership any of the above-mentioned titles or descriptions.

BYLAWS 12 AND 13: DUTIES OF MEMBERS

12

i. Each Member of the Institution of whatever class shall be bound to further to the best of his ability and judgement the objects, purposes, interests and influence of the Institution.

ii. Members of the Institution shall comply with the following Code of Conduct:

(a) Every member of whatever class shall so order his conduct as to uphold the dignity and reputation of the Institution and shall observe the provisions of the Charter and these Bylaws.

(b) Every Corporate Member shall at all times so order his conduct as to uphold the dignity and reputation of his profession and safeguard the public interest in matters of safety, health and otherwise. He shall exercise his professional skill and judgement to the best of his ability and discharge his professional responsibilities with integrity.

iii. The Code of Conduct set out in paragraph (ii)(b) of this Bylaw may from time to time be varied as allowed by the Lords of her Majesty's Most Honourable Privy Council.

13

i. Subject to paragraph (iii) of this Bylaw, the Council shall:

(a) From time to time make and publish Rules of Professional Conduct which shall be consistent with the Code set out in Bylaw 12 (ii) and shall require all members of whatever class to conform to such Rules.

(b) From time to time make and publish Disciplinary Regulations for the enforcement of the Code of Conduct, of Bylaw 12 and of the Rules of Professional Conduct made under paragraph (i)(a) of this Bylaw.

(c) Have the power to vary and rescind the Rules of Professional Conduct and Disciplinary Regulations referred to in paragraphs (i)(a) and (i)(b) respectively of this Bylaw as they think fit.

ii. The Disciplinary Regulations referred to in (i)(b) of this Bylaw shall *inter alia*:

(a) Include provision for the establishment, by and under the Authority of the Council, of:

A committee to investigate alleged breaches of the Code of Conduct of Bylaw 12 and of the associated rules established in (i)(a) of this Bylaw. A Committee to hear all charges of professional misconduct by a member and, as the occasion arises, an Appeals Committee whose decision shall be final.

(b) Confer upon the member the following rights at any hearing of the charges against him:

To attend in person and to be heard either in person with his solicitor or counsel or other representative.

To submit evidence in writing and to call witness in his defence.

To cross-examine witnesses other than those called by him.

To be given any other reasonable opportunity to defend himself against the charge or charges made.

iii. None of the Rules of Professional Conduct and none of the Disciplinary Regulations referred to in paragraphs (i)(a) and (i)(b) of this Bylaw, nor any rescission or variation thereof, shall be in any way repugnant to the Charter or these Bylaws.

Subject to the provisions of this Bylaw and Disciplinary Regulations referred to herein, if any Corporate or Non-Corporate Member shall refuse or wilfully neglect to comply with any of these Bylaws or with the Rules of Professional Conduct referred to in this Bylaw or shall have conducted himself in such a manner as in the opinion of the Council either shall have rendered him unfit to remain a Corporate or Non-Corporate Member of the Institution as the case may be or shall be injurious to the Institution, such Corporate or Non-Corporate Member may, by a resolution of the Council, be expelled, suspended, reprimanded, warned, or advised in such manner as the Council shall decide.

RULES OF PROFESSIONAL CONDUCT

The following Rules are made and published by the Council of the Institution as required by Bylaw 13 (i)(a) of the Bylaws of the Institution.

1. A Code of Professional Conduct designed to cover broad ethical principles must necessarily be drawn up in general terms. These Rules published by the Council indicate the manner in which members are required to conduct themselves in most situations. For situations not specifically encompassed by the Rules, members are required to order their conduct in accordance with the principle that, in any conflict between a member's personal interests and the fair and proper interests of the community, the member's duty to the community should prevail.

2. A member may be found guilty of improper conduct if in the opinion of the Council his conduct is in breach of the Rules of Professional Conduct hereafter set down.

3. A member, when discharging his professional duties:

(a) shall satisfy himself as to the extent of those duties and if in doubt obtain such clarification or confirmation as is necessary to satisfy himself as to their extent before entering upon them and shall not accept professional obligations which he believes he has not sufficient competence or authority to perform;

(b) shall accept due responsibility for all work done by him or under his direct supervision and shall take all reasonable steps to ensure that persons working under his authority are competent to carry out the tasks assigned to them and that they accept personal responsibility for work done under the authority delegated to them;

(c) shall, when called upon to give an opinion in his professional capacity and based on the facts disclosed to him, give an opinion that is objective and reliable to the best of his ability;

(d) shall, if his professional advice is not accepted, take all reasonable steps to ensure that the person overruling or neglecting his advice is aware of the possible danger which he believes may result from such overruling or neglect.

4. A member shall take all reasonable care in his work to minimise the risk of death, injury or ill-health to any person and of damage to property. In his work,

a member shall respect all laws and statutory regulations applicable to the design, operation and maintenance of chemical and processing plant. In addition, a member shall have due regard for the need to protect working and living environments and the need to ensure efficient use of natural raw materials and resources.

5. In respect of his professional relationships with employers, colleagues or clients, a member:

(a) shall not maliciously or recklessly injure or attempt to injure whether directly or indirectly the professional reputation of another;

(b) shall disclose to his client or employer all or any profits, benefits or interests he may have in the matter in which he is engaged;

(c) shall neither communicate to any person nor publish any information or matter, which was not previously known by him or published in the public domain, communicated to him in confidence by a client or employer without the express authority of that client or employer.

6. A member shall not improperly solicit work as an independent advisor or consultant either directly or by an agent, nor shall he improperly pay any person, by commission or otherwise, for the introduction of such work.

7. A member shall not make payment on his employer's behalf unless so requested by his employer nor shall he, in connection with work on which he is employed, place contracts or orders except with the authority of and on behalf of his employer.

8. A member, in connection with work in a country other than his own, shall order his conduct according to this Code so long as it is applicable: but where there are different standards of professional conduct in that country which are recognised by the Council he may adhere to them.

9. A member shall be prepared to take reasonable steps to further the education and training of candidates for the qualification of Chartered Engineer.

10. In respect of personal conduct a member shall at all times be mindful of the dignity of the profession. A member may be deemed guilty of improper conduct in the opinion of Council if:

(a) He is convicted by a competent tribunal of a criminal or civil offence: or,

(b) He is precluded from performing his professional duties in a manner which is consistent with the standards of his profession as a result of having been adjudicated bankrupt or of making a composition with his creditors: or,

(c) His conduct in other respects renders him unfit to be a member of the Institution or is injurious to the Institution.

APPENDIX B — TYPICAL CONSULTANCY SERVICES AGREEMENT

This AGREEMENT is made in (*insert place*) as of this (*insert day*) day of (*insert month*) 1994

BY AND BETWEEN (*insert name of owner*)
a company organized and existing under the laws of (*insert country*) and having its head office located at (*insert owner's address*) (hereinafter referred to as the 'OWNER')

AND (*insert consultant's name*)
a company organized and existing under the laws of England and having its head office located at (*insert consultant's address*) hereinafter referred to as the CONSULTANT.

WHEREAS
a. The OWNER desires to have provided by the CONSULTANT certain CONSULTANCY SERVICES, as hereinafter defined, in conncction with the (*insert type of plant*) which the OWNER is considering building in (*insert location*);

and

b. the CONSULTANT experienced in providing such services is willing to provide the services required by the OWNER.

WITNESSETH THAT in consideration of the covenants and agreements hereinafter set forth, the parties the signature of whose authorized officials are affixed hereto hereby agree as follows:

ARTICLE 1 — DEFINITIONS

Wherever used in this AGREEMENT, the following defined terms except as otherwise provided or otherwise required by the context shall have the meaning

assigned to them in Articles and Paragraphs referred to below:
AGREEMENT as defined in preamble.
OWNER as defined in preamble.
CONSULTANT as defined in preamble.
PROJECT as defined in Article 2 hereinafter.
CONSULTANCY SERVICES as defined in Paragraph 3.1 hereinafter.

ARTICLE 2 — PROJECT

The PROJECT for which the CONSULTANT shall furnish the CONSULTANCY SERVICES shall concern (*insert brief description of project*).

ARTICLE 3 — CONSULTANCY SERVICES

3.1 The CONSULTANT shall provide in connection with the PROJECT the CONSULTANCY SERVICES as defined in the Scope of Work document attached as ANNEX A and the Schedule attached as ANNEX B.

3.2 In order to perform the CONSULTANCY SERVICES, the OWNER will make available such information and assistance as may be necessary as defined in Annex C.

ARTICLE 4 — COMPENSATION TO CONSULTANT

4.1 As complete compensation for the performance of the CONSULTANCY SERVICES as defined in Article 3.1, the CONSULTANT shall be reimbursed the sum of (*insert fee*) Pounds Sterling (or if on schedule of rates refer to rates specified in place of fixed fee in this Clause).

4.2 The amounts under Article 4.1 shall be payable in Pounds Sterling at the following times against issue of the CONSULTANT'S invoice therefore: (percentages refer to the complete compensation specified in 4.1 above).
a. on authorization to proceed.
b. on issue of interim report to OWNER.
c. upon issue of the CONSULTANT'S final report on PROJECT.
(or such other payment terms as may be agreed).

4.3 Any delays in payment of the amounts due under Article 4.2 or 4.4 shall be subject to interest at 1% over the LIBOR (London Interbank Offer Rate) averaged over the period of delay.

4.4 Any work additional to the scope as defined in Article 3.1 shall be carried out under the following terms:
a. For each man-day expended by the CONSULTANT, the sum of (*insert fee*) Pounds Sterling.
b. For all travel, accommodation, incidental and out-of-pocket expenses relating to visits away from the CONSULTANT'S home office the CONSULTANT shall be reimbursed fully at cost by the OWNER.
c. Expenses incurred by the CONSULTANT for telephone (other than local calls), telex, telefax, courier or other communications shall be reimbursed at cost.
d. A man-day consists of seven and a half/eight hours. Time spent in travelling shall be counted as working time.
e. Payments under Article 4.4 shall be effected by the OWNER in Pounds Sterling within 30 days from the date of the CONSULTANT'S invoices.
f. The schedule under Article 3.1 shall be adjusted to reflect any additions to the scope of work requested by the OWNER and accepted by the CONSULTANT.

4.5 All payments shall be made to (*insert CONSULTANT'S Bank details*).

4.6 For the CONSULTANCY SERVICES performed by the CONSULTANT, all fees, rates and expenses specified in Article 4 are exclusive of taxes, duties, imposts or other charges wherever they may occur, other than UK Corporation Tax. The OWNER will indemnify the CONSULTANT against any aforesaid taxes, duties, imposts, etc other than Corporation Tax.

ARTICLE 5 — LIABILITY

The CONSULTANT'S sole and total liability under this AGREEMENT shall be to rectify errors and/or omissions in the CONSULTANT'S work which are shown to be the fault of the CONSULTANT. Other than the aforementioned the CONSULTANT shall not be liable for any direct or indirect or consequential damages arising from the PROJECT or the CONSULTANT'S CONSULTANCY SERVICES without limitations of any kind. Rectification under this

Article shall be invited to rework and/or redesign by the CONSULTANT as may be appropriate.

ARTICLE 6 — GOVERNING LAW

This AGREEMENT and the legal relationships of the parties thereto shall be construed, interpreted and governed by the law of England.

ARTICLE 7 — CONFIDENTIALITY

The CONSULTANT shall maintain confidential all correspondence, discussions, notes, reports, conclusions and/or results of the CONSULTANCY SERVICES with respect to any third parties and any persons not directly involved in the performance of the CONSULTANCY SERVICES.

The OWNER undertakes to maintain in confidence information received from the CONSULTANT for the purpose of this AGREEMENT and to use it only for the purpose of the PROJECT.

ARTICLE 8 — WHOLE AGREEMENT

This AGREEMENT and ANNEXES attached hereto shall constitute the whole AGREEMENT between the parties. No oral or written commitment by either party prior to the EFFECTIVE DATE defined hereunder shall have any legal effect on the parties.

ARTICLE 9 — EFFECTIVE DATE

This AGREEMENT shall come into full force and effect when the respective officers of the parties have affixed their names hereunder and the initial payment referred to in paragraph 4.2 has been transferred to the CONSULTANT'S account.

IN WITNESS WHEREOF the parties have caused this AGREEMENT to be executed, in their respective names by their respective officers, therefore duly authorized, as of the date and year first written above.

ANNEXE A

SCOPE OF WORK

(Insert name of Project)

(The scope of work is a detailed description of Consultant's activities as determined by the parties and covered by the Agreement.)

ANNEXE B

SCHEDULE OF CONSULTANT'S WORK

(*Insert name of Project*)

(Schedule of consultant's work

in appropriate format:

• milestones • bar chart • critical path)

ANNEXE C

SCHEDULE OF INFORMATION

REQUIRED TO BE PROVIDED BY OWNER

(*Insert name of Project*)

(List all information, documents, permissions, data etc. required to be provided by the OWNER for the CONSULTANT to carry out the work, together with dates by which they are to be made availble)

APPENDIX C — EXAMPLE OF TERMS AND CONDITIONS FOR WORK AS AN EXPERT WITNESS

EXPERT TECHNICAL ADVICE

Name of Expert

Chartered Chemical Engineer

Terms and Conditions of Contract

1. Fees and Costs

1.1 The fee is £** per hour. A timesheet will be maintained on which time spent on the Project will be logged in multiples of 5 minutes. In addition to the fee other expenses incurred will be charged on the basis shown below:

1.2 Travel:	
Public transport, taxis, hire cars, air fares etc (travel by train will be first class, air travel will be by business class).	at cost against receipts; (where practicable)
Car mileage:	**pence/mile
1.3 Hotel accommodation, meals etc:	at cost against receipts;
1.4 Telephone and fax:	at cost against invoices from BT and Mercury Communications.
All telephone/fax calls outside the London area (0171 and 0181) will be by Mercury and will be logged separately against a Project Code Number.	
1.5 Other agreed expenses, e.g. database searches, sampling, chemical analyses.	at cost against receipts; (where practicable)

2 VAT

VAT will be added to all invoices which are submitted to, and paid by, organizations registered and/or operating from within the UK.

3 Submission of invoices

Invoices will be submitted monthly together with timesheet and receipts.

4 Payment

4.1 Payment will be made within 28 calendar days of the date of the invoice and regardless of taxation (if any) by the courts and whether or not the client is seeking or obtaining Legal Aid.

4.2 Time of payment is of the essence and if there be any delay in payment interest may be added at a rate of 5% above the base lending rate and will accrue from day to day.

4.3 Payment shall be made in pounds sterling by means of a Banker's Draft or, in the case of a UK firm of solicitors or companies registered in the UK, by means of a firm's cheque.

4.4 Payment shall be in accordance with Principle 10.21 of the Law Society's *Guide to the Professional Conduct of Solicitors.*

5 Attending Court, Meetings or Conference, and Travelling

5.1 Fees will be charged at the rate specified in section 1 for time spent in attending meetings. Fees will be charged on an hourly basis calculated in multiples of 5 minutes.

5.2 The fee for attending Court will be £*** per day, or part thereof, and will become payable once the requirement for attendance is confirmed whether or not it later becomes unnecessary to attend the court for any, or part, of the time and irespective of the reason for such non-attendance.

6 Preparation of the Report and Presentation of Expert Evidence

6.1 The Report shall be prepared in accordance with the guidance notes of the British Academy of Experts and shall take account of the comments of Lords Denning and Wilberforce in Whitehouse v. Jordan in the Court of Appeal (1980) 1 All ER65 and the House of Lords (1981) 1 WLR 246 respectively (see Annexe 1). Expert evidence shall also be presented in accordance with the foregoing guidance and advice.

6.2 Unless otherwise agreed, the Report shall be first prepared in draft form so that the client, the instructing solicitor and counsel may have the opportunity to correct errors of fact which may inadverdently occur in the Report and to challenge the Expert on significant points in the Report. The final Report and Evidence as presented to the court shall be the Expert's exclusive responsibility.

7 Governing Law

The contract shall be governed by English Law and shall be subject to the exclusive juridiction of the English Courts in all matters.

Signed:...(Instructing solicitor)

Date:...

ANNEXE 1

Comment of Lord Denning in Whitehouse v. Jordan (1980) 1 All ER65:

'In the first place, their joint report suffers to my mind from the way it was prepared. It was the result of long conferences between two professors and counsel in London, and was actually "settled" by counsel. In short, it wears the colours of special pleadings rather than an impartial report. Wherever counsel "settle" documents we know how it goes. "We had better put this in", "We had better have this out", and so forth. A striking instance is the way in which Professor Tizard's was "doctored". The lawyers blocked out a couple of lines in which he agreed with Professor Strong that there was no negligence.'

Lord Wilberforce in the House of Lords (1981) 1 WLR 246:

'I have to say that I felt some concern as to the manner in which part of the expert evidence called for the plaintiff cause to be organized. The matter was discussed in the Court of Appeal and commented on by Lord Denning. While some degree of consultation between experts and legal advisers is entirely proper, it is necessary that expert evidence presented to the Court should be, and should be seen to be, the independent product of the expert, uninfluenced as to the form or content by the exigency of litigation. To that extent it is not, the evidence is likely to be not only incorrect but self-defeating. '

APPENDIX D — SO YOU WANT TO BE A CONSULTANT?

This Appendix is deliberately provocative. In tries to pose some of the 'awkward' questions which will arise in some form at some time during your consultancy career. There are no 'answers' suggested — only you can find answers within your own circumstances, personality and ambition. Seeing some of the questions in very plain language may enable you to plan how you will deal with the implications of these questions when they arise — as they will!

OBJECTIVES

- What do you want to achieve by being a consultant?
- Build up a full-time practice possibly involving partners or engaging staff?
- Develop a full-time one-man practice?
- Develop a committed but part-time one-man practice?
- Just fill in time and earn some money in addition?

THE MARKET

Achieving any of your objectives means that someone must buy your expertise. What is the market like? Ask yourself:

- What exactly is the market you are aiming for?
- How big is the market?
- Is the market local to your area, nationwide or international?
- What is the competition?
- Is the need for your expertise recognized in the market?
- Does the market use consultants?
- What benefits would using you give to a client?
- How else could the client obtain these benefits?
- Are the price levels in the market high enough to support a professional consultancy approach?
- Can you sell expertise only or must you also supply hardware, where extra financial consequences arise?

YOURSELF

Knowing your objectives and the market you are aiming at, what about yourself and your personal circumstances — are they consistent with achieving your objectives?

What exactly is your expertise?
- Can you work effectively entirely on your own?
- Can your skills be used effectively in a one-person environment?
- Can you pursue your goals alone and in the face of apathy or opposition?
- Are you a self-starter or do you need to be given a job definition?

Which of these are you?
- Technologist
- Manager
- Organizer
- Co-ordinator

Which of the above are you selling to your clients?

What is your spouse/family's attitude to your ambition? Are they:
- dedicated to its success?
- supportive?
- accepting it without involvement?
- apathetic?
- opposed to it?

And finally, and probably the most important, how good are you at coping with stress and uncertainty? Can you keep calm under pressure? There will be times when it looks as if you are going to run out of work and you will need to be able to cope with the thin periods as well as those periods when you are having to work seven days a week.

YOUR FUTURE

How many years of practice could you reasonably plan for before:
- you want to retire full-time?
- your skills become outdated?
- you would like to sell your practice?

How much time and energy must you spend in building and serving the practice you wish to achieve?
Can you put in the necessary hours per month regularly to sustain your practice?

RUNNING THE PRACTICE

As a one-man practice, you are entirely on your own. Facilities you may have taken for granted in your employment with an operating company now have to be provided by you. Typical facilities required by your practice will include:

TECHNICAL INFORMATION

- library
- literature searches
- catalogues
- magazines
- data banks

ADMINISTRATION

- typing
- secretarial
- copying
- printing
- archiving and filing
- office supplies and stationery including domestic services
- transport and travel services
- computing

COMMUNICATIONS

- mail
- telephone and answering service
- fax
- courier

FINANCIAL

- financial backing
- banking
- accountancy and book-keeping including VAT
- taxation and pensions

COMMERCIAL
- marketing and promotion
- sales
- contract administration

LEGAL
- liabilities — professional and public
- contract conditions
- patents and copyright

Some aspects of all these subjects must be covered, either by seeking (and paying for) professional services or by undertaking the tasks yourself.

Handling all these matters will necessarily occupy much non-earning and non-technical time which must be allowed for in your expectation of how much you can earn in the time you are able to devote to your practice.

APPENDIX E — FURTHER SOURCES OF ADVICE AND ASSISTANCE

OTHER ORGANIZATIONS

Confederation of British Industry
Centre Point
103 New Oxford Street
London WC1A 1DU
tel: 0171 379 7400
fax: 0171 240 1578

Council for Small Businesses in Rural Areas
141 Castle Street
Salisbury
Wilts SP1 3TP
tel: 01722 336255
fax: 01722 332769
Offers advice and consultancy services through the County Rural Industries Organizer.

Department of Trade and Industry
Admin HQ and Central Services
123 Victoria Street
London SW1E 6RB
tel: 0171 215 5000 (Enquiries)
fax: 0171 215 5665

Institute of Management
Management House
Cottingham Road
Corby
Northants NN17 1TT
tel: 01536 204222
fax: 01536 201651

International Bank for Reconstruction and Development (World Bank)
1818 H Street North West
Washington
DC 20433
USA
tel: + 1 (202) 477 1234
fax: + 1 (202) 477 6391
Maintains a computer-based register of experienced consultants. Applications to register must be made direct to Washington.

SERVICES FROM THE IChemE

INSURANCE BROKERS

A list of insurance brokers who provide professional indemnity insurance for consultant engineers is maintained by the Institution's Technical department and is available on request.

LIBRARY AND INFORMATION SERVICE

The library is open from 9.00 am to 5.00 pm, Monday to Friday and holds an extensive range of chemical engineering text books, journals and reference books. An on-line search service is also available. All enquiries should be directed to the Information Officer.

PUBLICATIONS

IChemE publishes a large number of books and training packages. These are available for purchase via the IChemE Book Sales Department at Rugby.

SETTING UP IN BUSINESS

This is a useful list of publications, government schemes and other sources of information which is available on request from IChemE's Engineering Department.